CAVALRY AND YEOMANRY BADGES OF THE BRITISH ARMY 1914

Other titles in the same series:

Badges of the British Army, 1820-1960 by F. Wilkinson
Buttons of the British Army, 1855-1970 by Howard Ripley
Campaign Medals of the British Army, 1815-1972 by Robert W. Gould
Old British Model Soldiers, 1893-1918 by L.W. Richards
Scottish Regimental Badges 1793-1971 by W.H. and K.D. Bloomer
Glengarry Badges of the British Line Regiments to 1881 by W.Y. Carman

CAVALRY AND YEOMANRY BADGES OF THE BRITISH ARMY 1914

BY F. WILKINSON

a&ap

ARMS AND ARMOUR PRESS

Published by
Arms and Armour Press
Lionel Leventhal Limited
2–6 Hampstead High Street
London NW3 1PR

SBN 85368 399 9

ACKNOWLEDGEMENTS
With few exceptions all the items shown are from the collection of Ron Pain whose assistance is gratefully acknowledged; he allowed his fine display to be disturbed as well as providing much valuable information about the badges. The photographs were taken, under somewhat difficult conditions, by Paul Forrester and Fred Pryor to whom sincere thanks are offered.

Printed in Great Britain
Produced by Dexin (Printers) Ltd. and
Duotech Graphics Ltd.

Contents

Introduction

For the collector of limited means badges have a popular appeal. Most are small, a virtue for those with limited space; they are usually attractive in appearance; they offer encouragement for collectors keen to carry out research; and finally they are still relatively inexpensive. One of the problems facing collectors is that of choice, for the variety of badges is enormous, each country having equipped its forces with many styles of badge in many materials, and each with its own special appeal. For the majority of British collectors it is hardly surprising that the badges of the British Army are most in demand.

The British army has used regimental badges from the eighteenth century but it was during the nineteenth century that they became more general. Metal badges were worn on the various headdresses, on the collars and sleeves of the tunic as well as on belts. Of all types cap badges are most sought after by modern collectors, but as the range is so great most try to specialise. One of the most popular groups are those of the mounted troops, both regular and yeomanry.

This volume is intended to cater for the collector who wants to specialise in the badges of the mounted troops of the British Army. It is possible to find an occasional badge of the late eighteenth or early nineteenth century but for most collectors only twentieth century examples are to be found. Since the period c.1900–14 was probably the age of maximum interest in cavalry it was decided to use as a central theme for this book the British cavalry of 1914. Each regiment appearing in the Army List of August 1914 is represented in this volume, but to assist collectors in identification many badges worn before and after the Great War have been included.

Cavalry and Yeomanry

During the first half of the nineteenth century there were numerous changes in the organisation and uniforms of the British cavalry. A variety of metal helmets were worn and many of these were fitted on the brow with a large plate which generally carried the Royal cypher and the name of the regiment. Light dragoons were fitted out with a stiff shako, belling at the top, to the front of which was fitted a metal badge bearing the Royal cypher and regimental number. In 1805 some of the light dragoons were converted into hussar regiments and for a while they wore a copy of the original Hungarian hussar fur hat, or busby, and these were devoid of badges. Later the busby was replaced by a shako but in 1841 Hussar regiments were again wearing busbies.

Other Light Dragoon units were converted to lancer regiments and these also adopted the cap of the original Polish lancers. The lance cap (tschapka) consisted of a dome-shaped skullpiece with a truncated and inverted pyramidal top-piece and on the brow of the skullpiece was secured a large plate. The pattern varied but most had a royal motif and the regiment's name as well as any battle honours. Troopers' plates were stamped out of brass, whilst the officers had the applied decorations made separately, often of silver.

By the 1890s the British cavalry establishment consisted of five regiments of heavy cavalry, thirteen of medium cavalry, including lancers and dragoon guards, and 13 hussar regiments. All carried a sword and a carbine with 30 rounds. Warrant officers, certain sergeants and the trumpeters were issued with a Webley revolver and 12 rounds.

Backing up the regular cavalry were the volunteer troops, known collectively as the yeomanry. Although a few groups were founded earlier, the majority of units could trace their origins back to the Napoleonic Wars. In 1804 an act set out their conditions of service which were to stay largely unchanged for the rest of their lifetime namely, that they were to be called out in case of an invasion or to aid the civil powers. In 1888 they were made liable for service anywhere in Great Britain, but during the Boer War large numbers volunteered for service in South Africa.

Since the yeomanry were all volunteers and were therefore able to chose their own role, some opted to be hussars, some dragoons and others simply mounted riflemen. Their uniforms were basically similar to those of the appropriate branch of the regular cavalry but there were, of course, numerous variations in detail and colouring.

Dress uniforms were highly attractive but not always practical until many small campaigns and the Crimean War (1854–6) showed that simpler uniforms were essential for active service. Khaki replaced reds and golds, and simple caps the helmets, so that the large helmet plates previously worn were no longer appropriate for campaign wear; as a result the smaller cap badges of the type illustrated in this book came into being.

Since each of the yeomanry units was based on one locality it was not uncommon for their badges to incorporate parts of the county arms, or those of a member of the county nobility. In the case of regular units the emphasis was more on the Royal cypher and regimental number or device, or an award for some special action. Smaller versions of the badge, sometimes varying in detail, were worn on the lapels or collar of the tunic. In addition to those of the regiment, certain other badges were worn by NCOs on the sleeve of their tunic; sometimes they indicated rank and sometimes trade. The badges were fitted to the cap by one of two systems, either by a narrow bar sliding into a slot on the cap, or by two pierced lugs which passed through the material and were then secured by a split pin. Nearly all the regiments listed below were in existence in 1914 but following the Great War there was considerable pruning and reorganisation with the result that many regiments were amalgamated. This process has continued and is continuing even now.

Collecting Badges

Acquiring examples of badges is very much a matter of persistence, devotion and luck. The higher prices resulting from an increased demand has meant that antique dealers are a little more ready to search out and handle badges. A number of specialist dealers issue regular lists and more and more auction rooms are holding sales of militaria of all types. It is therefore possible for new collectors to build up a good selection of badges from these and other sources.

Cleaning is largely a matter of moderation and it must be remembered that every single polish removes a small amount of metal so that, given time, detail will be worn away. Some collectors like to lacquer the cleaned badge but this is a matter of personal preference.

Mounting is something of a problem since it is almost certain that rearrangements will become necessary as new specimens are acquired. Some collectors keep the badges in plastic envelopes but the majority like to display the collection and probably the simplest method is to mount the items on card. If a coloured card of medium thickness is used the badges can be laid out and then the appropriate slits cut to accommodate the slide or rings. It is advisable not to use very thick card or the forcing in of the slide may fracture the joint.

Labelling is a matter of personal choice and some collectors prefer to use a number only on the board and record full details on a card or in a book; others fix a stick-on label to the display board giving the title of the regiment and date.

A final word of warning is called for, as a new development has taken place over the past year or so. A number of extremely good reproduction badges have been produced and it is possible that more will appear. So far the method of securing the badges, theraded brass fitting, is a good indication, but if cap badges are produced then it will become necessary to look very carefully at every badge.

The Plates

To assist in easy identification the badges are reproduced full size and the details of the metal used given as well as the means of attachment on the particular badge illustrated (although this does not mean that examples using alternative methods were not used). The dates shown are those when authority was given for the wearing of that particular design but in many cases it is known that earlier patterns continued to be worn for some time after this date.

From 1915 to 1919, owing to the demands of war, many badges previously made in bi-metal format were produced in an all brass version.

The metal coding for the badges shown consists of the following abbreviations:

b brass
wm white metal
bi part brass, part white metal
g gilt
bz bronze

The method of securing the badges is indicated by 's' for slide, 'r' for rings and 't' for tabs.

Bibliography

Booth, E. Talbot, *The British Army*, London, 1940.
Carman, W. Y., *British Military Uniforms from Contemporary Pictures*, London, 1968.
Headdresses of the British Army: Cavalry, London, 1968.
Headdresses of the British Army: Yeomanry, London,1970.
Chichester, H., and Short, G. Burgess, *Records and Badges of The British Army*, latest edition, London, 1970.
Cooper, L., *British Regular Cavalry*, London, 1965.
Daniel, W. H., *Military Forces of the Crown*, London, 1901.
Edwards, T. J., *Regimental Badges*, latest edition, London, 1968.
Frederick, J. B. M., *Lineage Book of the British Army*, New York, 1969.
Farmer, J., *Regimental Records of the British Army*, London, 1901.
Gaylor, J., *Military Badge Collecting*, London, 1971.
King, E. Cooper, *The British Army and Auxiliary Forces*, two volumes, London, 1893.
Perry, O. L., *Rank and Badges*, London, 1871.
Rogers, H. C. B., *Mounted Troops of the British Army, 1066–1954*, London, 1959.
Simkin, R. and Archer, L. *British Yeomanry Uniforms*, London, 1971.

LIFE GUARDS

1

2

SECOND LIFE GUARDS

3

HONI SOIT QUI MAL Y PENSE

5

ROYAL HORSE GUARDS

4

KING'S DRAGOON GUARDS

6

7

K
DG

8

K
DG

9

10

11

12

CAVALRY BADGES

1. The 1st Life Guards; b; s.
2. The 1st Life Guards; b; tabs.
3. The 2nd Life Guards; b; s.
4. Royal Horse Guards (The Blues); b; s.
5. Household Battalion; b; s; c.1914.
6. The 1st (King's) Dragoon Guards; b; s; pre-1915 (when the Austrian eagle was discarded).
7. The 1st (King's) Dragoon Guards; b; s (No scroll).
8. The 1st (King's) Dragoon Guards; b; s; 1915–19.
9. The 1st (King's) Dragoon Guards; wm with b centre; s; 1915–37.
10. The 1st (King's) Dragoon Guards; wm; r; post-1937.
11. The 2nd Dragoon Guards (Queen's Bays); b; r; pre-1902 (as it bears the Victorian Crown).
12. The 2nd Dragoon Guards (Queen's Bays); b; s; post-1902 (Amalgamated with the 1st (King's) Dragoon Guards in 1959 to form the 1st The Queen's Dragoon Guards.)

13

14

15

16

17

18

19

20

21

22

23

24

13. The 3rd (Prince of Wales's) Dragoon Guards; wm with scroll and crown; g; r. (Amalagamated with the Carabiniers (6th Dragoon Guards) in 1922 to form 3rd Carabiniers (Prince of Wales's Dragoon Guards).

14. The 4th (Royal Irish) Dragoon Guards; wm with b scroll; s. (Amalgamated with the 7th Dragoon Guards (Princess Royal's) in 1922 to form the 4th/7th Royal Dragoon Guards.

15. The 5th (Princess Charlotte of Wales's) Dragoon Guards; b with wm centre; r; pre-1902.

16. The 5th (Princess Charlotte of Wales's) Dragoon Guards, b with wm centre; s; post-1902. (Amalgamated with the Inniskillings (6th Dragoons) in 1922 to form the 5th Royal Inniskilling Dragoon Guards.)

17. The 6th Regiment of Dragoon Guards (Carabiniers); b; r; pre-1902.

18. The 6th Regiment of Dragoon Guards (Carabiniers); b with wm centre and scroll; s; post-1902. (Amalgamated in 1922 with the 3rd Dragoon Guards.)

19. The 7th (The Princess Royal's) Dragoon Guards; wm; r; 1898–1906.

20. The 7th (The Princess Royal's) Dragoon Guards; b; s; post-1906.

21. The 1st (Royal) Dragoons; b with wm scroll; 1898–1902.

22. The 1st (Royal) Dragoons; brass with wm scroll; s; 1902–15 and 1919–48.

23. The 1st (Royal) Dragoons; gilt with wm wreath; s; 1948–69.

24. The 1st (Royal) Dragoons; a; s; c. 1915–19.

25

26

27

28

29

30

31

32

33

34

35

36

25. The 2nd Dragoons (Royal Scots Greys); wm; s.

26. The 2nd Dragoons (Royal Scots Greys); wm with g scroll.

27. Plume-holder grenade for bearskin cap of Royal Scots Greys; fitted with long slide.

28. White metal horse fitted at the back of the Scots Greys bearskin cap; it resembles S B5 but is easily recognised by the very big loops (½in long) needed to pierce the thick cap.

29. The 3rd (King's Own) Hussars; wm with g scroll; s.

30. The 3rd (King's Own) Hussars; g; post-1902. (Amalgamated with the 7th Queen's Own Hussars in 1958 to form The Queen's Own Hussars.)

31. The 4th (Queen's Own) Hussars; b with wm numerals; s; pre-1902.

32. The 4th, (The Queen's Own) Hussars; b with wm letters and scroll; r; post-1906. (Amalgamated in 1958 with the 8th King's Royal Irish Hussars to form The Queen's Royal Irish Hussars.)

33. The 4th (Queen's Own) Hussars; brass with wm white metal metal rings; post-1902 (no scroll).

34. The 5th (Royal Irish) Lancers; brass with wm figure and lower half of flags. (Amalgamated in 1922 with the 16th The Queen's Lancers to form The 16th/5th The Queen's Royal Lancers.)

35. The 6th (Inniskilling) Dragoons; wm with b scroll; s.

36. The 6th (Inniskilling) Dragoons; b; r (Note that the flag is in the opposite direction to the previous badge. Regiment amalgamated in 1922 with the 5th Dragoon Guards.)

37

38

39

40

41

42

43

44

45

46

47

37. The 7th (Queen's Own) Hussars; b with wm letters; 1898–1902.

38. The 7th (Queen's Own) Hussars; b with wm letters; r; post-1902. (Amalgamated in 1958 with the 3rd King's Own Hussars.)

39. The 7th (Queen's Own) Hussars; b; s; c.1916.

40. The 8th (The King's Royal Irish) Hussars; wm with g scroll and crown; r; 1898–1902.

41. The 8th (The King's Royal Irish) Hussars; wm with b crown and scroll; s; post-1902. (Amalgamated in 1958 with the 4th Queen's Own Hussars.)

42. The 9th (or Queen's Royal) Lancers; wm; r; 1898–1902.

43. The 9th (or Queen's Royal) Lancers; wm; s; 1902–53.

44. The 10th, The Prince of Wales's Own Royal Hussars; wm; g crown and scroll; s; 1914. (Amalgamated with 11th Hussars (Prince Albert's Own) in 1969 to form The Royal Hussars (Prince of Wales's Own.)

45. The 10th, The Prince of Wales's Own Royal Hussars; b; s; c.1916–1919.

46. The 11th Prince Albert's Own Hussars; g; s; 1914.

47. The 12th (Prince of Wales's Royal) Lancers; wm and g; r; 1898–1903.

48

49

50

51

52

53

54

55

56

57

58

59

48. The 12th (Prince of Wales's Royal) Lancers; wm and g; w; 1903–58. (Amalgamated with the 9th Queen's Royal Lancers in 1960.)

49. The 13th Hussars; b with wm numerals; r; 1898–1902.

50. The 13th Hussars; b with wm numerals; s; post-1902.

51. The 13th Hussars; b; r. (Worn on forage cap. Regiment amalgamated with the 18th Royal Hussars. (Queen Mary's Own) in 1922 to form the 13th/18th Royal Hussars (Queen Mary's Own).

52. The 14th (King's) Hussars; b with silver eagle; s; 1898–1915.

53. The 14th (King's) Hussars; b; s; 1915–29.

54. The 14th (King's) Hussars; b; s; 1915–29. (Amalgamated in 1922 with the 20th Hussars to form the 14th/20th Hussars.)

55. The 15th (King's) Hussars; b with wm crown and lion; r; pre-1902.

56. The 15th (King's) Hussars; b and wm; s. (Amalgamated with the 19th Royal Hussars (Queen Alexandra's Own) Hussars in 1922 to form the 15th/19th The King's Royal Hussars.)

57. The 16th, (The Queen's) Lancers; wm and b; r; pre-1902.

58. The 16th (The Queen's) Lancers; wm and b; s; c.1905–22. (Amalgamated with the 5th Royal Irish Lancers in 1922.)

59. The 17th Lancers (Duke of Cambridge's Own); wm; s. (Amalgamated with the 21st Lancers (The Empress of India's) in 1922 to form the 17th/21st Lancers.)

60

61

62

63

64

65

66

67

68

69

70

71

60. The 18th Hussars; b with wm numerals; r; 1898–1902.

61. The 18th (Queen Mary's Own) Hussars; wm; s; post-1911. (Amalgamated with the 13th Hussars in 1922.)

62. The 19th Hussars; wm; s; 1898–1902.

63. The 19th (Queen Alexandra's Own Royal) Hussars; wm; r; 1902–9 (double scroll).

64. The 19th (Queen Alexandra's Own Royal) Hussars; wm; s; 1909–22. (Amalgamated with the 15th The King's Hussars in 1922 to form the 15th/19th The King's Royal Hussars.)

65. The 20th Hussars; b; r; 1898–1902.

66. The 20th Hussars; b; s; 1902. (Amalgamated with the 14th Hussars in 1922.)

67. The 21st (Empress of India's) Lancers; b; lower flag wm; r; 1898–1902.

68. The 21st (Empress of India's) Lancers; b; s; c.1916–18, the same pattern with wm parls to flag 1902-1922. (Amalgamated with 17th Lancers in 1922.)

69. The 22nd Dragoons; wm; s; 1940–48.

70. The 23rd Hussars; wm with b crown; s; 1940–8.

71. The 24th Lancers; wm; r; 1940–8.

72

73

74

75

76

78

77

79

80

72. The 25th Dragoons; b with wm sword; s; 1941–8.

73. The 26th Hussars; b; s; 1941–8.

74. The 27th Lancers; b with silver elephant; s; 1940–8.

75. Dragoon Guards; b; r; 1900–2.

76. Dragoons; b; r; 1900–2.

77. Lancers; b with wm on lower part of flags; r; 1900–2.

78. The Life Guards; b; r; non voided centre formed by amalgamation the 1st and 2nd Life Guards in 1922.

79. The 3rd Carabiniers (Prince of Wales's Dragoon Guards); wm with b crown and scroll; r. (Formed by amalgamating the Carabiniers (6th Dragoon Guards) and the 3rd Dragoon Guards (Prince of Wales's) in 1922.)

80. The 4th/7th Royal Dragoon Guards; wm; s. (Formed by amalgamating the 4th Royal Irish Dragoon Guards and the 7th Dragoon Guards (Princess Royal's) in 1922.)

81. The 5th Royal Inniskilling Dragoon Guards; wm; s. (Formed in 1922 by amalgamating The Inniskilling (6th Dragoons) and the 5th Dragoon Guards (Princess Charlotte of Wales's), but this title was only awarded in 1935.)

82. The 13th/18th Royal Hussars (Queen Mary's Own); b; s.

83. The 13th/18th Royal Hussars (Queen Mary's Own) b with wm letters; r. (Formed by amalgamating the 13th Hussars and the 18th Royal Hussars (Queen Mary's Own) in 1922.)

84. The 14th/20th King's Hussars; b; s. (Formed in 1922 by amalgamating the 14th King's Hussars and the 20th Hussars.)

81

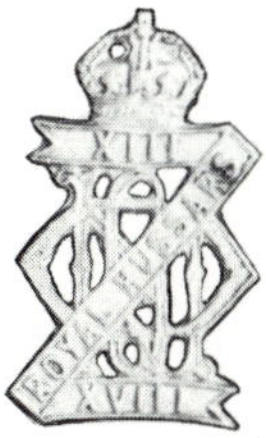

82

83

84

85

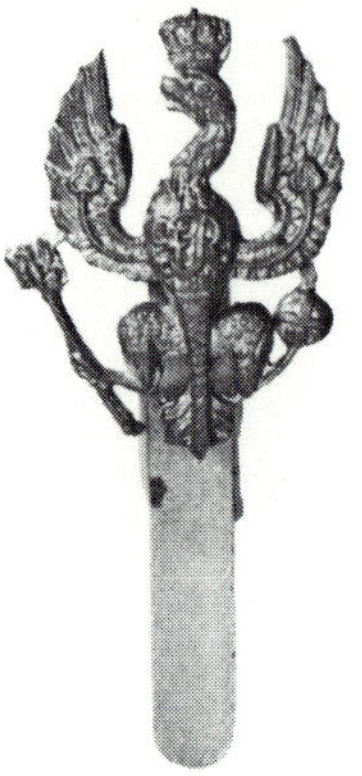

86

87

88

89

90

91

92

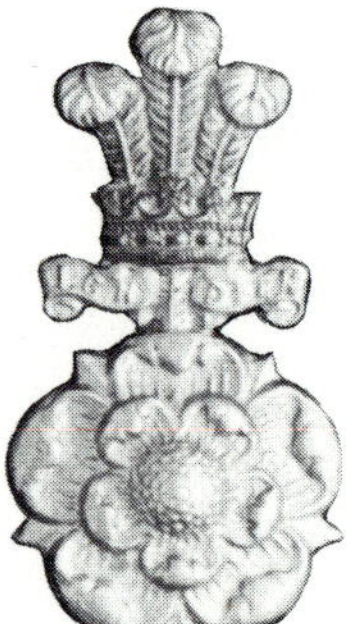

93

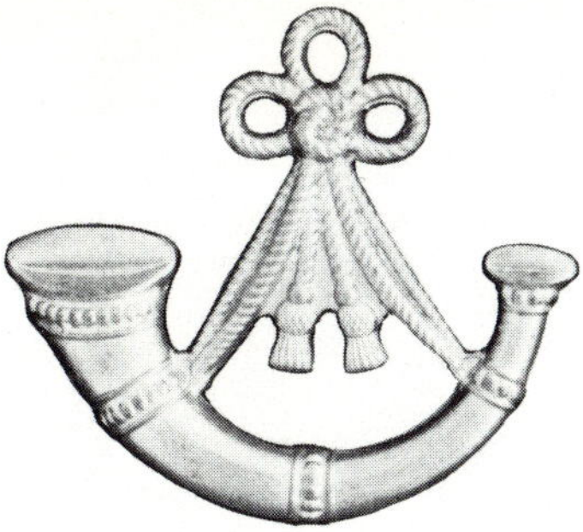

94

95

96

85. The 14th/20th King's Hussars; b; r; c.1929–31.

86. The 14th/20th King's Hussars; b; a; letters 'FR' on chest of eagle.

87. The 15th/19th The King's Royal Hussars; b with crown and lion in wm; s. (Formed by amalgamating The 15th The King's Hussars and the 19th Royal Hussars (Queen Alexandra's Own) in 1922.)

88. Army Remount Service; b with horse in wm; s; c.1914.

YEOMANRY BADGES

89. The Royal Wiltshire Yeomanry (Prince of Wales's Own); b coronet with wm feathers; r.

The Warwickshire Yeomanry:
90. Imperial Yeomanry, wm,r.
91. b; r.
92. wm; r.

93. The Yorkshire Hussars (Alexandra, Princess of Wales's Own); wm with g coronet; s.

The Nottinghamshire Yeomanry:
94. b; s; pre-1949.
95. (Sherwood Rangers); b; r; 1949–53.

The Staffordshire Yeomanry:
96. Imperial Yeomanry; wm; r.

97

98

99

100

101

102

103

104

105

106

107

108

The Staffordshire Yeomanry (continued):
97. NCO sleeve badge; wm; r; pre-1902.
98. b; s; pre-1902.
99. b; r; pre-1902.
100. b; s; post-1902.

The Shropshire Yeomanry:
101. b; s; post-1902 (see also
102. b; s; post-1950.

103. The Shropshire Yeomanry; b; r

The Ayrshire (Earl of Carrick's Own) Yeomanry:
104. b; s; pre-1915.
105. b; r; post-1923.

The Cheshire (Earl of Chester's) Yeomanry:
106. Imperial Yeomanry; wm; s; pre-1908.
107. b; s.
108. wm with b crown and scroll; s; post-1908.

109

110

111

112

113

114

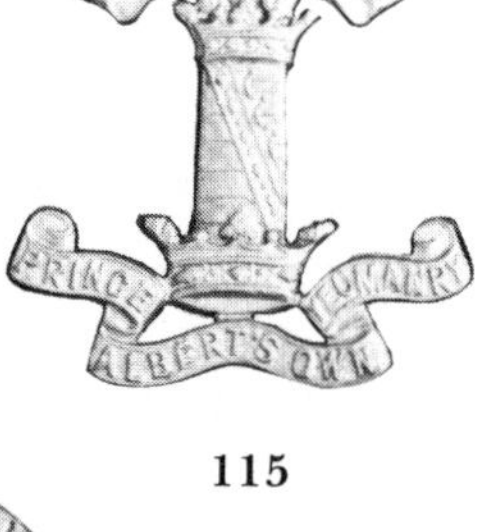

115

116

117

118

119

120

The Cheshire Yeomanry (continued):

109. b; s; post-1908.

The Yorkshire Dragoons:

110. sleeve badge; wm; r; pre-1902.
111. bz; s; 1914.
112. wm; s.
113. bz; s.

The Leicestershire Yeomanry (Prince Albert's Own):

114. b; s; 1908–15.
115. b; r; 1915–22.
116. b; s; scroll for honours; 1922–66.
117. wm; r; small badge worn on beret.

The North Somerset Yeomanry:

118. wm; r; Edward VII cypher.
119. wm; s; George V cypher.

The Duke of Lancaster's Own Yeomanry:

120. wm; pre-1902.

121

122

123

124

125

126

127

128

129

130

131

The Duke of Lancaster's Own Yeomanry (continued):

121. b; s; 1902–54 (see also
122. b with wm centre; s; 1954–66

123. The Lanarkshire Imperial Yeomanry; b; r; pre-1908.

The Northumberland Hussars:

124. b; r.
125. wm; r; pre-1902.
126. wm; s; c.1950.

The South Nottinghamshire Hussars:

127. b; r.
128. Officers' b; s.
129. wm; s.

130. The Denbighshire Hussars; b; s.

131. The Westmorland and Cumberland Yeomanry; b; s; 1902–c.1920.

132

133

134

135

136

137

138

139

140

141

142

143

132. The Pembroke Yeomanry (Castlemartin); wm with g coronet; r.

The Royal East Kent Yeomanry:
133 b; s; 1902–20.
134. bz; s.

The Hampshire Yeomanry:
135. b with wm rose and scroll; s.
136. b; s; 1907–20.

The Royal Bucks Hussars:
137. wm; s; post-1908.
138. b; r.

The Derbyshire Yeomanry:
139. Imperial Yeomanry b; r; 1902–8.
140. b; s.
141. wm; s.

The Dorsetshire Yeomanry:
142. Imperial Yeomanry b; r; 1902–8.
143. b; r; 1908–20.

144

145

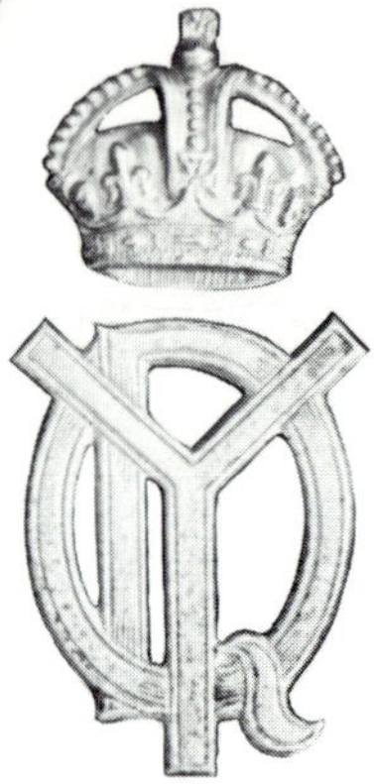

146

147

148

149

150

151

152

153

154

155

156

The Dorsetshire Yeomanry (continued):

144. b; s; 1920–60.
145. wm; s; 1920–60.
146. wm; r.

The Royal Gloucestershire Hussars:

147. Imperial Yeomanry; b; r; 1902–8.
148. b; r; post-1908.

149. The Hertfordshire Yeomanry; b; r.

The Berkshire Yeomanry:

150. Imperial Yeomanry; wm; r.
151. b; r.

The 1st County of London Yeomanry (Middlesex, The Duke of Cambridge's Hussars):

152. wm; w; Edward VII cypher.
153. bz; s; George V cypher.
154. b; s; George V cypher.
155. b; r; George V cypher with scroll.
156. wm; r; George VI cypher.

157

158

159

160

161

162

163

164

165

166

167

157. The 1st Royal Devon Yeomanry Hussars; bz; r; 1902–20.

The Duke of York's Own Loyal Suffolk Hussars
158. wm; t.
159. Brass with scroll in wm; s.

160. The Royal North Devonshire Hussars; b; r; pre-1920.
161. b; s.

The Queen's Own Worcestershire Hussars:
162. b; s; post-1908.
163. wm; s; post-1908.

Royal Devon Yeomanry Artillery:
164. b; s; post-1924.
165. bz; s; post-1924.

The West Kent Yeomanry (Queen's Own); wm; s.
166. wm; s.

167. The West Somerset Yeomanry; b; s.

168

169

170

171

172

173

174

175

176

177

178

179

180

The Queen's Own Oxfordshire Hussars:

168. wm; r.
169. bz; r.
170. wm; s.

The Montgomeryshire Yeomanry:

171. wm; s; pre-1908.
172. Imperial Yeomanry; b; s; pre-1908.
173. wm; s; post-1908–18.

174. The Lothians and Border Horse; b; s; pre-1908.

The Lothians and Berwickshire Imperial Yeomanry:

175. b; s; pre-1908.
176. b; s; 1908-56.

The Queen's Own Royal Glasgow Yeomanry:

177. Brass with rings 1902–c.1920.
178. b; r; small badge worn on cap.

The Lanarkshire Yeomanry:

179. b; r.
180. b; r; post c.1953.

181

182

183

184

185

186

187

188

The Lancashire Hussars:
181. b; r.
182. Imperial Yeomanry b; r; pre-1908.

183. The Lancashire Hussars; bz; r; 1908–20.

184. The Surrey Imperial Yeomanry; wm; r; 1901–10.

The Queen Mary's Regiment Surrey Yeomanry:
185. wm; r; post-1910.
186. b; s.

The Fife and Forfarshire Yeomanry:
187. b; s; 1908–56.
188. wm; s; 1908–56.

The Norfolk Yeomanry (The King's Own Royal Regiment):
189. b; r; 1936–53.
190. b; r; 1902–8.
191. b; r; 1902–8. (probably a collar dog).

192. The Sussex Yeomanry; b; s; post-1908.

189

190

191

192

193

194

195

196

197

198

199

200

193. The Glamorgan Yeomanry; wm with g scroll and crown; s.

194. The Lincolnshire Yeomanry; b; s; post-1908.

The City of London:
195. b; r; figure in wm.
196. bz; s.
197. Imperial Yeomanry; wm arms with blue centre.
198. b; s; wm centre.
199. 'R R' brass, spurs wm; r; 1947–50.
200. b; wm centre and scroll; s.

The 2nd County of London Yeomanry (Westminster Dragoons):
201. Imperial Yeomanry b; r; pre-1908.
202. wm; r; 1908–38.
203. b; w; 1908–38.

The 3rd County of London:
204. Imperial Yeomanry; b; r; 1901–10.

201

202

203

204

205

206

207

208

209

210

211

The 3rd County of London (continued):

205. b; r; 1901–10.
206. b; s; 1910-20.
207. (23rd); brass with wm numerals; in wm; 1920–38.
208. brass with wm letters; 1938–9.
209. brass; with slide, letters in wm; 1939–44.

210. The Bedfordshire Yeomanry; b; r; 1914.

211. NCO's arm badge; wm; r; 1914.

The Essex Yeomanry:

212. b; r; 1901–5.
213. b; r; 1905–9.
214. b; r; 1905–9.
215. b; r; 1909–16.
216. b; s; 1918–54.

212

213

214

215

216

217

218

219

220

221

222

223

224

225

North of Ireland Imperial Yeomanry:

217. wm with rings; 1902–8.
218. Brass with rings.

North Irish Horse:

219. wm with rings; 1952–53.
220. brass with rings; 1908–c.1952.

The Northamptonshire Yeomanry:

221. wm; s; post-1908.
222. 2nd Regiment; wm; r.

The East Riding Yeomanry:

223. b with wm scroll; r 1920–56.
224. b with FORWARD in wm; r; 1920–56.

225. The East Riding Yeomanry; b; r; c.1914–18.

Lovat's Scouts

226. b; r; pre-1908.
227. b; r; post-1908.
228. wm; r; 1903–22.
229. wm; r; 1900–03.

226

227

228

229

230

231

232

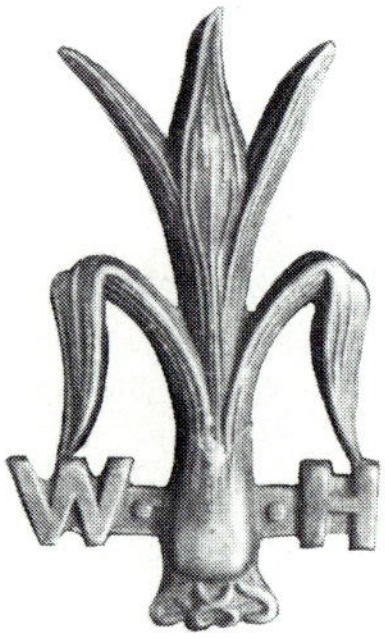

233

234

235

236

237

238

239

240

Scottish Horse

230. wm; r; 1900–20.
231. b; r; 1900–20.
232. wm; r; post-1902.

Welsh Horse Yeomanry

233. Brass with slide. Raised 1914 – amalgamated in 1917 with Montgomeryshire Yeomanry.
234. Imperial Yeomanry Rosette for slouch hat; letters b on mauve and red rosette.

Auxiliary Yeomanry Units

235. King Edward's Horse (The King's Overseas Dominions Regiment). Brass with rings 1910–24.
236. King Edward's Horse (The King's Overseas Dominions Regiment). wm with slide.
237. 2nd King Edward's Horse (The King's Overseas Dominions Regiment). Brass with slide. 1914–18.

South Irish Horse

238. b; r; pre-1908.
239. b; r; 1908–22.

240. League of Frontiersmen (A body of ex-servicemen). Brass with slide.

241

242

243

244

245

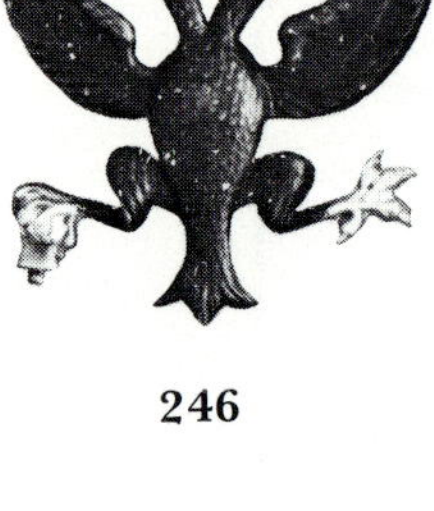

246

247

248

249

250

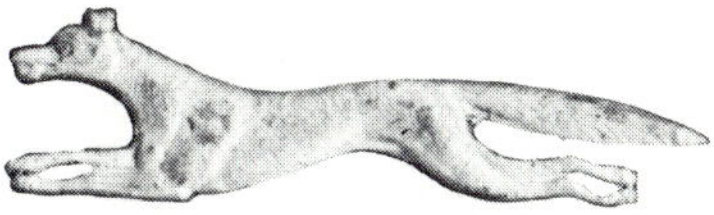

251

252

241. British American Squadron of King's Colonials, Imperial Yeomanry. 1901–1905. Brass with rings.

OFFICERS' BADGES

Officers' Badges (unless otherwise stated, all are fitted with tabs):

242. Royal Wiltshire Yeomanry, silver and gilt; r.
243. Warwickshire Yeomanry; bz.
244. North Somerset Yeomanry; silver and enamel.
245. North Somerset Yeomanry; bz.
246. Lanarkshire Yeomanry; black and gilt; r.
247. Royal Bucks Hussars; bz; r.
248. Royal North Devon Hussars; bz.
249. Lancashire Hussars Imperial Yeomanry; bz.
250. Sussex Yeomanry; bz; pre-1906.
251. East Riding of Yorkshire Imperial Yeomanry; g.
252. Middlesex Imperial Yeomanry, wm.

253

254

255

256

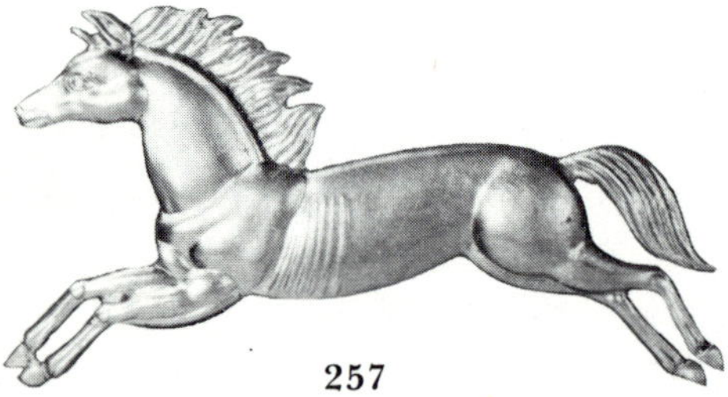

257

258

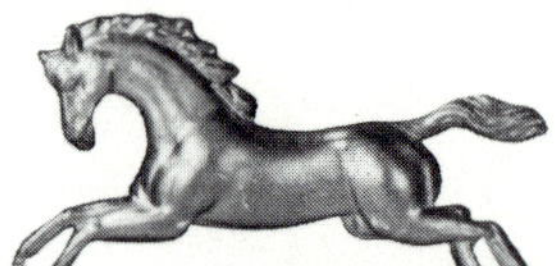

259

260

261

262

SLEEVE BADGES

(In many of the cavalry regiments senior NCOs wear a special badge on their right arm above the chevrons. All are of white metal and attached by rings unless otherwise stated.)

253. 1st (King's) Dragoon Guards.

254. 2nd Dragoon Guards (Queen's Bays).

255. 6th Dragoon Guards (Carabiniers); crown b.

256. 4th/7th Royal Dragoon Guards.

257. 5th Dragoon Guards.

258. 2nd Dragoons (Royal Scots Greys).

259. 3rd (King's Own) Hussars.

260. 7th (Queen's Own) Hussars; t.

261. 8th (King's Royal Irish) Hussars; pre-1902.

262. 10th (Prince of Wales's Own Royal) Hussars.

14/20 H.
Junior NCOs Chrome
Senior NCOs Black/Gilt.

263

264

265

266

267

268

269

270

COLLAR BADGES

263. 17th Lancers; b; 1.2in across.

264. 9th Queen's Royal Lancers; 1953–60; 1.3in across.

265. 18th Hussars; 1901; wm figures, wreath b; 1.3in across.

266. 18th (Queen Mary's Own) Hussars; post-1902; wm; 1in across.

267. 13th Hussars; 1901–22; wm figures; b; 1in across.

268. 7th Queen's Own Hussars; 1901–58; b with wm monogram; 0.8in across.

269. Fife and Forfar Yeomanry 1901–56; b; 1in across.

270. 5th Royal Inniskilling Dragoon Guards; castle wm; scroll b; service dress badge; 0.8in across.

271. 11th Hussars (Prince Albert's Own); b.

272. 4th Queen's Own Hussars; 1906–58; figures and scroll wm, rest b; 0.8in across.

273. The Royal Dragoons (1st Dragoons); wreath and tablet wm, rest b; 1in across.

271

272

273

274

275

276

277

278

279

280

281

282

283

284

285

286

BUTTONS

274. 5th Dragoon Guards (Princess Charlotte of Wales's) 1855–1922; g; by Jennens & Co., London; 0.9in diameter.

275. 1st (Royal) Dragoons; 1855–1904; g; by Jennens & Co., London; 0.9in diameter.

276. The Royal Dragoons (1st Dragoons); 1904–69; b; by Smith & Wright Ltd., Birmingham; 0.9in diameter.

277. 6th Dragoon Guards; 1881–1920; b; by Jennens & Co., London; 0.9in diameter.

278. 9th Queens Royal Lancers; 1855–1901; b; by Jennens & Co., London; 0.9in diameter.

279. 4th/7th Royal Dragoon Guards; post-1922; b; by J.R. Gaunt & Son, London; 1in diameter.

280. Sometimes confused with 281 is this button of the Royal Irish Fusiliers (Princess Victoria's) but close examination will show the difference.

281. The Royal Scots Greys (2nd Dragoons); post-1877; b; by Firmin & Sons, London; 0.9in diameter.

282. The Leicestershire Yeomanry (Prince Albert's Own); pre-1956; b; by J.R. Gaunt & Son Ltd., London; 0.9in diameter.

283. Scottish Horse; 1902; b; by Wm. Anderson & Sons, Edinburgh; 1in diameter.

284. Lincolnshire Imperial Yeomanry; 1901–8; wm; by Jennens & Co., London; 0.9in diameter.

285. 18th Light Dragoons; ball button by Jennens & Co., London; wm; 0.5in diameter.

286. Insignia worn in addition to stripes by Collarmakers in the Royal Artillery; b; 1.4in x 1.6in.

Appendix

CAVALRY REGIMENTS, AUGUST 1914 (with facings and station)

1st Life Guards: Uniform scarlet. Facings blue. Plume white. Station Hyde Park.
2nd Life Guards: Uniform scarlet. Facings blue. Plume white. Station Windsor.
Royal Horse Guards (The Blues): Uniform blue. Facings scarlet. Plume red. Station Regent's Park.
1st (King's) Dragoon Guards: Uniform scarlet. Facings blue. Plume red. Station Dunbar.
2nd Dragoon Guards (Queen's Bays): Uniform scarlet. Facings buff. Plume black. Station Newport, Mon.
3rd (Prince of Wales's) Dragoon Guards: Uniform scarlet. Facings yellow. Plume black and red. Station Newport, Mon.
4th (Royal Irish) Dragoon Guards: Uniform scarlet. Facings blue. Plume white. Station Newport, Mon.
5th (Princess Charlotte of Wales's) Dragoon Guards: Uniform scarlet. Facings dark green. Plume red and white. Station Dunbar.
6th Dragoon Guards (Carabiniers): Uniform blue. Facings white. Plume white. Station Newport, Mon.
7th (Princess Royal's) Dragoon Guards: Uniform scarlet. Facings black. Plume black and white. Station Newport, Mon.
1st (Royal) Dragoons: Uniform scarlet. Facings blue. Plume black. Station Dunbar.
2nd Dragoons (Royal Scots Greys): Uniform scarlet. Facings blue. Plume white. Station Dunbar.
3rd (King's Own) Hussars: Uniform blue. Collars scarlet. Busby-Bag. Garter blue. Plume white. Station Bristol.
4th (Queen's Own) Hussars: Uniform blue. Busby-Bag yellow. Plume scarlet. Station Dublin.
5th (Royal Irish) Lancers: Uniform blue. Facings scarlet. Plume green. Station Woolwich.
6th (Iniskilling) Dragoons: Uniforms scarlet. Facings primrose. Plume white. Station Newport, Mon.
7th (Queen's Own) Hussars: Uniform blue. Busby-Bag scarlet. Plume white Station Bristol.
8th (King's Royal Irish) Hussars: Uniform blue. Busby-Bag scarlet. Plume red and white. Station Dublin.
9th (Queen's Royal) Lancers: Uniform blue. Facings scarlet. Plume black and white. Station Woolwich.
10th (Prince of Wales's Own Royal) Hussars: Uniform blue. Busby-Bag scarlet. Plume black and white. Station Scarborough.
11th (Prince Albert's Own) Hussars: Uniform blue. Overalls crimson. Busby-Bag crimson. Plume crimson and white. Station Dublin.
12th (Prince of Wales's Royal) Lancers: Uniform blue. Facings scarlet. Plume scarlet. Station Woolwich.
13th Hussars: Uniform blue. Collars buff. Busby-Bag buff. Plume white. Station Dublin.
14th (King's) Hussars: Uniform blue. Busby-Bag yellow. Plume white. Station Scarborough.
15th (The King's) Hussars: Uniform blue. Busby-Bag scarlet. Plume scarlet. Station Bristol.
16th (The Queen's) Lancers: Uniform scarlet. Facings blue. Plume black. Station Woolwich.
17th (Duke of Cambridge's Own) Lancers: Uniform blue. Facings white. Plume white. Station Woolwich.
18th (Queen Mary's Own) Hussars: Uniform blue. Busby-Bag blue. Plume scarlet and white. Station Scarborough.
19th (Queen Alexandra's Own Royal) Hussars: Uniform blue. Busby-Bag white. Plume white. Station Bristol.
20th Hussars: Uniform blue. Busby-Bag crimson. Plume yellow. Station Scarborough.
21st (Empress of India's) Lancers: Uniform blue. Facings French grey. Plume white. Station Woolwich.

YEOMANRY REGIMENTS, AUGUST 1914 (with facings and station)

Special Reserve

North Irish Horse: Uniform green. Facings white. Plume green. Station Skegoniel Avenue Belfast.

South Irish Horse: Uniform green. Facings red and green. Station Beggars Bush Barracks Dublin.

King Edward's Horse (The King's Overseas Dominions Regiment): Uniform drab. Facings scarlet. Plume black. Station Duke of York's H.Q. Chelsea SW.

Ayrshire (Earl of Carrick's Own): Uniform blue. Facings and Busby-Bag scarlet. Plume scarlet and white. Station Ayr.

Bedfordshire: Uniform blue. Facings white. Plume black and white. Station Ashburnham Rd., Bedford.

Berks (Hungerford): Uniform blue. Facings scarlet. Station Reading.

Buckinghamshire (Royal Bucks Hussars): Uniform green. Facings and Busby-Bag scarlet. Plume white. Station Buckingham.

Cheshire (Earl of Chester's): Uniform blue. Facings scarlet. Busby-Bag white. Plume red and white. Station Old Bank Building, Chester.

Denbighshire (Hussars): Uniform blue. Facings and Busby-Bag scarlet. Plume white. Station 1, Erdigg Road, Wrexham.

Derbyshire: Uniform blue. Facings scarlet. Plume red and white. Station 91, Siddall's Rd. Derby.

Royal 1st Devon: Uniform scarlet. Facings blue. Plume scarlet and white. Busby-Bag scarlet. 9, Dix's Field, Exeter.

Royal North Devon (Hussars): Uniform blue. Facings and Busby-Bag scarlet. Plume scarlet and white. Station Barnstaple.

Dorset (Queen's Own): Uniform blue. Facings and Busby-Bag scarlet. Plume white. Station Sherborne.

Essex: Uniform green. Facings scarlet. Plume scarlet. Station 17, St. Isaac's Walk, Colchester.

Fife and Forfar: Uniform scarlet. Facings blue. Station Kirkcaldy.

Glamorgan: Uniform blue. Facings and Plume white. Station Bridgend.

Gloucestershire (Royal Gloucestershire Hussars): Uniform blue. Facings blue. Busby-Bag scarlet. Plume scarlet and white. Station The Barracks, Gloucester.

Hampshire (Carabiniers): Uniform blue. Facings and Plume white. Station Hyde Close, Winchester.

Herts: Uniform scarlet. Facings white. Plume black. Station Hertford.

Royal East Kent (The Duke of Connaught's Own) (Mounted Rifles): Uniform rifle green. Facings scarlet. Plume red and green. Station Canterbury.

West Kent (Queen's Own): Uniform blue. Facings and Busby-Bag scarlet. Plume red and white. Station Drill Hall, Union St. Maidstone.

Lanarkshire: Uniform blue. Facings scarlet. Station Lanark.

Lanarkshire (Queen's Own Royal Glasgow and Lower Ward of Lanarkshire): Uniform dark blue. Facings scarlet. Plume black. Station Yorkhill Parade, Yorkhill, Glasgow.

Lancashire Hussars: Uniform blue. Busby-Bag crimson. Plume crimson and white. Station Prince Alfred Rd. Liverpool.

Duke of Lancaster's Own: Uniform scarlet. Facings blue. Plume white. Station Lancaster House, Whalley Road, Whalley Range, Manchester.

Leicestershire (Prince Albert's Own): Uniform blue. Facings scarlet. Busby-Bag red. Plume white. Station Leicester.

Lincolnshire: Uniform green. Facings white. Plume green. Station Old Barracks, Lincoln.

City of London (Rough Riders): Uniform blue grey. Facings purple. Plume light blue. Station 39, Finsbury Square E.C.

1st County of London (Middlesex Duke of Cambridge's Hussars): Uniform green. Facings black. Busby-Bag scarlet. Plume green and scarlet. Station Duke of York's Headquarters, Chelsea, S.W.

2nd County of London (Westminster Dragoons): Uniform scarlet. Facings purple. Plume white. Station Elverton St. Westminster S.W.

3rd County of London (Sharpshooters): Uniform green. Facings and Busby-Bag green. Plume white. Station Henry St. St. John's Wood, N.W.

Lothians and Border Horse: Uniform scarlet. Facings blue. Plume white. Station 7, Wemyss Place, Edinburgh.

1st Lovat's Scouts: Uniform blue. Facings blue. Station Beauly N.B.

2nd Lovat's Scouts: Uniform blue. Facings blue. Station Beauly N.B.

Montgomeryshire: Uniform scarlet. Facings black. Plume white. Station Welshpool.

Norfolk (The King's Own Royal Regiment): Uniform blue. Facings yellow. Plume yellow. Station Cattle Market St. Norwich.
Northamptonshire: Uniform blue. Facings light blue. Plume light blue and white. Station Clare St. Northampton
Northumberland (Hussars): Uniform blue. Busby-Bag scarlet. Plume scarlet and white. Station Northumberland Road, Newcastle-on-Tyne.
Nottinghamshire (Sherwood Rangers): Uniform green. Facings green. Busby-Bag scarlet. Plume green and white. Station Retford.
Nottinghamshire (South Nottinghamshire Hussars): Uniform blue. Busby-Bag scarlet. Plume red and white. Station Derby Road, Nottingham.
Oxfordshire (Queen's Own Oxfordshire Hussars): Uniform dark blue. Facings and Busby-Bag mantua purple. Plume mantua purple and white. Station Oxford.
Pembroke (Castlemartin): Uniform blue. Facings white. Station Tenby.
Scottish Horse: Uniform Atholl green. Facings yellow. Plume black cock feathers. Station Dunkeld N.B.
Shropshire: Uniform blue. Facings scarlet. Plume red and white. Station Shrewsbury.
North Somerset: Uniform blue. Facings and Plume white. Station Bath.
West Somerset: Uniform blue. Facings scarlet. Busby-Bag red. Plume white. Station County Territorial Hall, Taunton.
Staffordshire (Queen's Own Royal Regt): Uniform blue. Facings and Busby-Bag scarlet. Plume white. Station Barley St., Stafford.
Suffolk (The Duke of York's Own Loyal Suffolk Hussars): Uniform green. Facings and Busby-Bag scarlet. Plume white. Station Bury St. Edmunds.
Surrey (Queen Mary's Regiment): Uniform blue. Facings blue. Station Melbourne House, King's Avenue, Clapham Park S.W.
Sussex: Uniform blue. Facings yellow. Station Drill Hall, Church St. Brighton.
Warwickshire: Uniform dark blue. Facings, Busby-Bag and Plume white. Station St. John's Warwick.
Westmorland and Cumberland: Uniform scarlet. Facings white. Busby-Bag scarlet. Plume red and white. Station Penrith.
Royal Wiltshire (Prince of Wales's Own Royal Regiment): Uniform blue. Facings scarlet. Station The Butts, London Road, Chippenham.
Worcestershire (The Queen's Own Worcestershire Hussars): Uniform blue. Facings and Busby-Bag and Plume scarlet. Station Worcester.
Yorkshire Dragoons (Queen's Own): Uniform blue. Facings and Plume white. Station Doncaster.
Yorkshire Hussars (Alexandra, Princess of Wales's Own): Uniform blue. Busby-Bag scarlet. Plume black and scarlet. Station York.
East Riding of Yorkshire: Uniform maroon. Facings light blue. Plume light blue and white. Station Railway Street, Beverley.

ORDER OF PRECEDENCE OF THE REGIMENTS OF YEOMANRY

1. Royal Wiltshire.
2. Warwickshire.
3. Yorkshire Hussars.
4. Nottinghamshire (Sherwood Rangers).
5. Staffordshire.
6. Shropshire.
7. Ayrshire.
8. Cheshire.
9. Yorkshire Dragoons.
10. Leicestershire.
11. North Somerset.
12. Duke of Lancaster's Own.
13. Lanarkshire.
14. Northumberland.
15. Nottinghamshire Hussars South.
16. Denbighshire.
17. Westmorland & Cumberland.
18. Pembroke.
19. Royal East Kent
20. Hampshire.
21. Buckinghamshire.
22. Derbyshire.
23. Dorset.
24. Gloucestershire.
25. Herts.
26. Berks.
27. 1st County of London.
28. Royal 1st Devon.
29. Suffolk.
30. Royal North Devon.
31. Worcestershire.
32. West Kent.
33. West Somerset.
34. Oxfordshire.
35. Montgomeryshire.
36. Lothians & Border Horse.
37. Lanarkshire (Glasgow).
38. Lancashire Hussars.
39. Surrey.
40. Fife & Forfar.
41. Norfolk.
42. Sussex.
43. Glamorgan.
44. Lincolnshire.
45. City of London.
46. 2nd County of London.
47. 3rd County of London.
48. Bedfordshire.
49. Essex.
50. Northamptonshire.
51. East Riding of Yorkshire.
52. 1st Lovat's Scouts.
53. 2nd Lovat's Scouts.
54. Scottish Horse.

Index